Clothes

Clothes

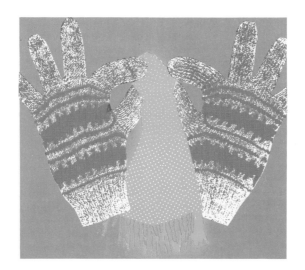

Heather Govier

GEC GARRETT EDUCATIONAL CORPORATION

Edited by Richard Young

Text © 1991 by Garrett Educational Corporation First Published in the United States in 1991 by Garrett Educational Corporation, 130 East 13th Street, Ada, Oklahoma 74820

First Published in 1987 by Macdonald & Co. (Publishers) Ltd., London with the title CLOTHES © 1987 Macdonald & Co. (Publishers) Ltd.

All rights reserved including the right of reproduction in whole or in part in any form without the prior written permission of the publisher. Published by Garrett Educational Corporation, 130 East 13th Street, Ada, Oklahoma 74820

Manufactured in the United States of America.

Library of Congress Cataloging - in - Publication Data
Govier, Heather.
    Clothes / Heather Govier.
       p. cm. — (First technology library)
    "First published in 1987 by Macdonald & Co. (Publishers) Ltd., London" — T. p. verso.
    Summary: Discusses clothing and how it is manufactured and distributed. Includes suggestions for related projects.
    ISBN 1-56074-009-4
    1. Clothing and dress — Juvenile literature. 2. Clothing trade — Juvenile literature. [1. Clothing and dress. 2. Clothing trade.]
  I. Title II. Series
TT497.G68   1991
687 — dc20   91-20533
                                  CIP
                                  AC

How to use this book

First, look at the Contents page opposite. Read the list to see if it includes the subject you want. The list tells you what each page is about, so you can find the page with the information you need.

In the book, some words are **darker** than others. These are harder words. Sometimes there is a picture to explain the word. For example, the words **chain stitch** appear on page 20, and there is a picture of one on the same page. Other words are explained in the Word List on page 31.

On page 28 you will find a technology project. This project suggests ideas and starting points for learning more about technology by yourself.

CONTENTS

WHAT ARE CLOTHES? — 6-7
Why wear them? — 6-7

IN THE BEGINNING — 8-15
Using nature — 8-9
Changing nature — 10-11
Spinning thread — 12-13
Dyeing and bleaching — 14-15

MAKING CLOTHES — 16-21
Weaving and knitting — 16-17
Cutting out — 18-19
Sewing up — 20-21

FROM SHOP TO YOU — 22-27
At the factory — 22-23
Doing the shopping — 24-25
Looking after clothes — 26-27

TECHNOLOGY TO TRY — 28-30
Keeping clothes clean — 28-30

WORD LIST — 31

WHAT ARE CLOTHES?

Why wear them?

What are your favorite clothes? Perhaps you like to wear a warm-up suit or a pair of shorts and a T-shirt. Or perhaps you enjoy dressing up in your best clothes to go to a party.

All over the world people wear different kinds of clothes. Snug sweaters and coats help to keep us warm, and loose cotton clothes keep us cool when it is very hot. Even when the weather is just right, most people like to wear some clothes because they would feel silly with nothing on at all.

Sometimes you wear special clothes just to show that you belong to a group. If all your friends wear jeans, then you probably feel more comfortable when you are wearing jeans too. Uniforms are one special way of showing that we belong to a group. Do you wear a uniform for school? How many other kinds of uniforms can you think of?

Do you like fancy dress parties? It is fun to wear clothes to draw attention to yourself and to make you look good.

People need to wear different clothes for work depending on what they do. Those with dirty jobs wear overalls or aprons, firemen wear special heatproof suits, and people working at a building site need to wear **safety helmets** in case something falls on their heads.

IN THE BEGINNING

Using nature

The first clothes were made from animal skins. Even in the **Stone Age** people learned how to sew skins together with needles made from bone. We still wear animal skins today. Fur or sheepskin coats are very warm for the winter, and many different kinds of clothes are made from leather.

Some people think that it is cruel to kill animals for their skins and that we could manage just as well without wearing animal skins at all.

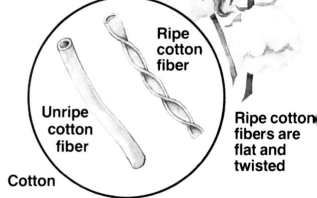

Cotton

Unripe cotton fiber

Ripe cotton fiber

Ripe cotton fibers are flat and twisted

Most of our clothes today are not made from animal skins but from sheets of fabric. These are made from **yarns** of twisted **fibers** of different kinds. Natural fibers come from plants or from animals. Cotton is a fiber that comes from the seed of the cotton plant. It grows in hot countries of the world, such as Egypt, India and parts of the United States. Cotton makes a fabric that is cool to wear when the weather is hot.

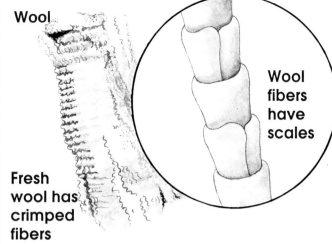

Wool

Fresh wool has crimped fibers

Wool fibers have scales

The most common animal fiber is wool, which comes from sheep. There is no need to kill the sheep to take its wool. It is just sheared off, rather like giving the sheep a haircut. Clothes made from wool can be very warm and snug to wear when it is cold. Silk and flax are other types of natural fibers. Can you find out where they both come from?

Stone Age people sewed animal skins

Making a needle from bone

The bone is broken with a flint and then sharpened

Changing nature

Cotton and wool are **fibers** that grow on plants or animals. They have been used to make clothes for thousands of years.

In the past 100 years people have found out how to make new kinds of fibers. These are **manufactured** fibers. They do not grow but are made in factories from oil, coal or wood. Many of the clothes we wear today are made from these fibers.

Manufactured fibers have some advantages over natural fibers. Cotton or wool clothes are very comfortable but hard to keep clean. Woolen clothes shrink if washed in very hot water and cotton clothes need a lot of drying. Cotton clothes also need careful ironing to keep them looking smart and crisp. Clothes made from manufactured fibers are easy to wash and many of them are quick drying, which means that they need less ironing. Fabrics made from a mixture of natural and manufactured fibers can have the good points of both.

Manufactured fibers are made by treating oil, coal or wood with various **chemicals** until they turn into thick, sticky liquids. The liquids are then squeezed through very small holes, forming fine threads that harden and are chopped into fibers. These manufactured fibers can be made into clothes in just the same way as natural fibers.

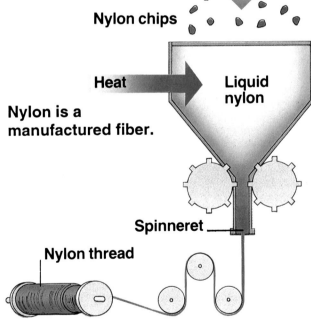

Nylon is a manufactured fiber.

10

TRY THIS

Clothes to keep us warm

Can you find out what keeps us warmest? Fill some squeeze bottles with hot water and wrap each one up in a different cloth. Use a thermometer to check the temperature of the water every 15 minutes. Which bottle stays hot the longest? Try the test again using other cloths. Does it matter what color they are?

Oil → nylon, polyester, acrylic
Coal ⏋
Wood → viscose, acetate

Manufactured fibres

Polyester and viscose

Nylon

Acrylic

Polyester

Acrylic

Acetate and nylon

Nylon

Spinning thread

Fibers are like short hairs. They are only a few inches long and too short to make into cloth. First the fibers must be spun together to make one long thread.

People have known how to spin for thousands of years. Cotton or wool fibers start off in clumps. These are slowly pulled out then stretched and twisted to make a long thread. This is wound into a ball or onto a **bobbin.**

Spinning factory about 1840

At first the twisting and pulling was done by hand with a weight such as a stone or a spindle tied to the end of the yarn. The invention of the spinning wheel made the job much faster.

People first used spinning wheels in their own homes. But when huge machines were invented to do the spinning, people had to go to work in large factories. Inside the early spinning mills, it was noisy and dirty. Many people were needed to keep the machines running. They worked long hours. Today, spinning factories are better places to work—clean and quiet—and many of the machines are controlled by computers.

People did spinning by hand for thousands of years

Spindle

DID YOU KNOW?

Some historical facts

People have known how to spin for 4,000 years. The first spinning wheel was invented in the 14th century. At that time, it took a week to spin about six pounds of wool. Then, in 1779, Samuel Compton invented a machine called a **mule.** With a mule, each machine operator could spin as much in one day as a person used to spin in one year.

Spinning wheel about 200 years ago

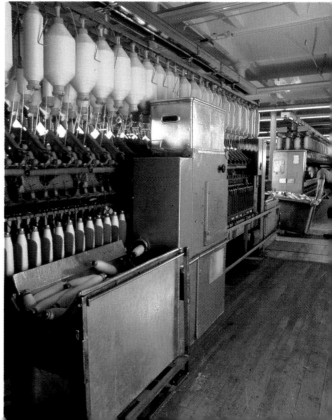

Modern spinning factory

Dyeing and bleaching

Cotton and wool are a yellowish white color when they are first washed and dried. Through the ages, people have used **dyes** to make their clothes more colorful.

Simple **dyes** can be made from many plants that grow wild. Think how blackberries stain your hands purple when you pick them.

To make cloth really shining white, we bleach it with **chemicals.** In olden times, the only way to do this was to leave the cloth out in the sun.

Bleaching cloth about 150 years ago

Modern dyes made from oil give bright colors

Today, chemicals are used for bleaching and for dyeing. Dyes made from plants do not give very bright colors, but modern chemical dyes made from oil can be very bright and all the colors of the rainbow. Another problem with natural dyes is that after a time the colors fade or wash out. Chemical dyes last much longer and can be sealed into the fabrics so that they will not run out and stain other clothes that are washed with them.

Sometimes the dyes are used on the yarn itself, but often the fabric is dyed after the yarn has been woven into cloth. By putting the colors on afterwards, patterns can be made. Modern dyes and modern techniques mean that our clothes can be more colorful and more interesting than ever before.

Tie dyeing

Make a simple dye from onion skins or blackberries and use it to dye a white handkerchief. Tie knots in the handkerchief, put it in a pan of water with the blackberries or onion skins and boil them for about half an hour. **Get an adult to help you with this.** Rinse the handkerchief in clean water until no more color comes out. When the handkerchief is completely dry, untie the knots. Do you like the pattern you have made?

Crimson from madder

Brown from onion

Blue from woad

Until 130 years ago, people made dyes from plants to color their clothes

Yellow from weld

Modern dye factory

MAKING CLOTHES

Weaving and knitting

Do you know someone who likes to do a lot of knitting? Knitting is one way of making clothes from balls of wool. Because all you need are some large needles, you can easily do knitting in your own home.

There are many other simple ways of making yarn into cloth, such as **crochet, tatting** and **lace-making.** People have done them for hundreds of years and are still doing them today.

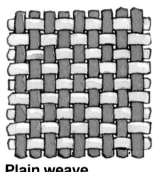
Plain weave

Moving from the country to the weaving towns changed people's lives
A weaving town in England in 1885

Twill weave

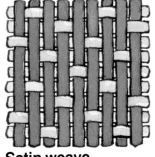
Satin weave

Most cloth is made from yarn by weaving or knitting it. At first, weaving was done at home by using simple looms that people built for themselves. These worked well, but it took a long time to make the cloth.

People have used simple hand looms to weave cloth for centuries

Modern knitting machine

At the start of the **Industrial Revolution,** huge machines were invented that could weave cloth much faster. Much more cloth was made, and it was sold to people all over the world. The weaving mills changed the way that people lived. They left their farms and moved to new towns to work in the weaving mills. Their whole lives changed completely.

Modern machines controlled by computers can weave or knit fancy patterns into the fabric. Knitting machines also make socks, stockings and all kinds of sweaters.

Hand knitting is still very popular

Cutting out

Before we can wear the cloth, it has to be made into clothes.

First the cloth has to be cut into pieces that are the right shape to fit our bodies. To make a T-shirt, you need a piece of cloth for the back, a piece for the front and two small pieces for the sleeves. You use a pattern made of paper to get the shape right. You pin each piece of paper to the cloth and then cut around the paper.

Cutting out at the factory

A different pattern is used for every piece of a garment, and most clothes have many more pieces than a simple T-shirt. Because cloth costs money, it is important to arrange the patterns carefully so that not too much fabric is wasted. Laying out the patterns so that there is very little waste was once a very slow, skilled job, but now computers help to do this more quickly.

When the pattern has been arranged on the cloth, the pieces are carefully cut out around the pattern. This can be done with scissors, but scissors can only cut out one garment at a time. In modern factories, people cut the fabric with long, sharp knives. This way, hundreds of the same dress or shirt can be made at once.

Modern machines work out the best way to lay out the patterns

Cutting out at home

Sewing up

Can you sew? Sewing by hand may be the oldest craft of all. Even during the **Stone Age,** people used bone needles to sew skins together to make clothes.

People have invented all sorts of different stitches to sew with. Some are very strong, like the **running stitch.** Some are very pretty, like the **chain stitch.** Some stitches are good for making hems or edges because they only show on one side of the cloth.

Sewing up for the clothes trade

Blanket stitch

Running stitch

Chain stitch

Sewing by machine is much faster than sewing by hand. Have you seen a sewing machine? It works by using two strands of thread, one on the top through the eye of the needle and the other underneath on the **shuttle.** As the needle goes up and down, the two threads are wrapped together in the middle of the fabric.

The clothes we buy in stores are sewn together in large factories or small workshops. Sometimes the pieces are sent out to people in their homes to be sewn together. The people who do this often work very hard for very little money. After the pieces have been sewn together, the fastenings must be stitched on. Zippers and buttons are the most common types of fasteners. Can you think of others?

TRY THIS

Making clothes strong

Get a piece of polyethylene, jersey and a piece of shirt or dress fabric. Cut them into strips 2 inches wide and 20 inches long. Pin one end of each strip to a desk. Put pins 2 inches from the end and in the middle of the strip. Hang weights on the other end until the strip tears. What kinds of fabric take the most weight? Now try gluing or sewing two strips of each fabric together. Test for strength.

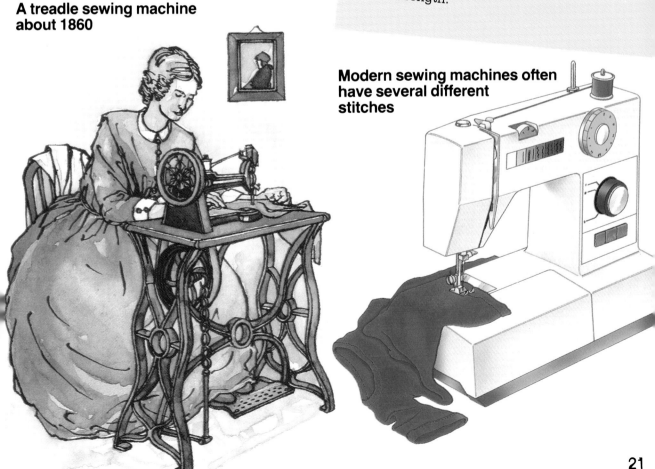

A treadle sewing machine about 1860

Modern sewing machines often have several different stitches

FROM SHOP TO YOU
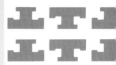

At the factory

Before new clothes leave the factory, someone checks them carefully. The person puts the clothes onto a frame and looks at the sewing and the material. They also make sure the right parts were sewn together.

The good clothes then go to the ironing room. Sometimes big machines do the work, but sometimes people use irons just like you have at home.

Afterwards, the clothes go on hangers inside polyethylene covers, or into polyethylene bags or boxes.

Computers keep a check on every **garment** that is made. All the bags and boxes have bar codes on them. A **bar code** is a pattern of black lines that can be read by a computer. The code tells the computer all about the clothes, such as what each garment is, what **fiber** it is made from, what size it is, and how much it costs. At the factory the computer counts all the garments that are sent to stores and sends out the bills to make sure that everything is paid for. The computer also records which clothes are selling best so that the factory can make more in time to meet orders from the stores.

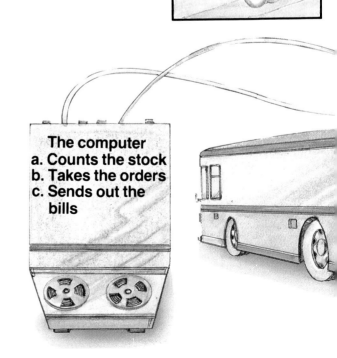

The computer
a. Counts the stock
b. Takes the orders
c. Sends out the bills

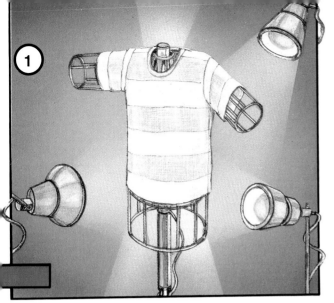

DID YOU KNOW?

More about bar codes

Here is a bar code. At the bottom is a number. This number is another way of writing the message in the bar code. The bar code tells the computer where the garment comes from, who the manufacturer is and the size and color of the garment.

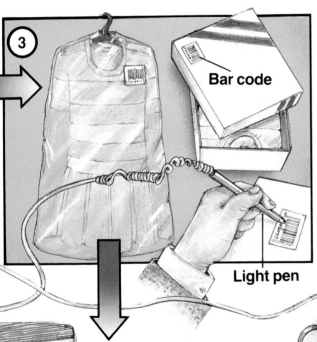

1. Looking for mistakes
2. Ironing by hand or by machine
3. Packing in bags and boxes
4. Sending the clothes to the shops
5. Checking the stock and taking orders

Doing the shopping

Catalog

Most stores now use computers to keep a check on all the clothes that people buy. After you pick out some new clothes, you take them to the cashier. The **cashier** finds the **bar code** on the clothes and passes a **light pen** over it. The light pen tells the computer inside the cash register to print out your **receipt** and the cashier takes your money.

The computer also adds up how many clothes are sold. It sends a message to the factory to order more clothes when they are needed.

Ordering clothes by electronic mail

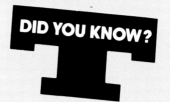

Some facts

Thousands of people work in the clothing industry in hundreds of factories throughout the United States. Only a third of all the clothes sold today are made from natural fibers. All the rest are made from manufactured fibers. In the United States, people spend billions of dollars on clothes every year.

You can buy clothes through mail order without going to stores. **Catalogs** are sent to homes with pictures of all the goods that are for sale. All you have to do is fill in a form telling what you want to buy and pop it in the mail. An even quicker way is to order clothes by telephone. After placing your order, the clothes will be packed into a parcel and will be sent directly to your home.

People who have computers at home sometimes can order clothes by electronic mail. That is, they type their order into their computer and then send it by telephone to a computer at a **warehouse.** They may even type in their bank account number so that they pay for the clothes by electronic mail as well.

Looking after clothes

If you want your clothes to look good, you must hang them up when you take them off so that they don't get all wrinkled. Most clothes have a label inside to tell us what they are made of and how to take care of them. Can you find such a label in any of your clothes?

When your clothes become dirty, they must be washed. This is much easier today because we have **automatic** washing machines to do the job. In olden times, everything was washed by hand. Because the dirty clothes needed lots of scrubbing to get them clean, it was very hard work. Drying and ironing clothes are also easier now that we have **tumble driers** and electric steam irons to help us.

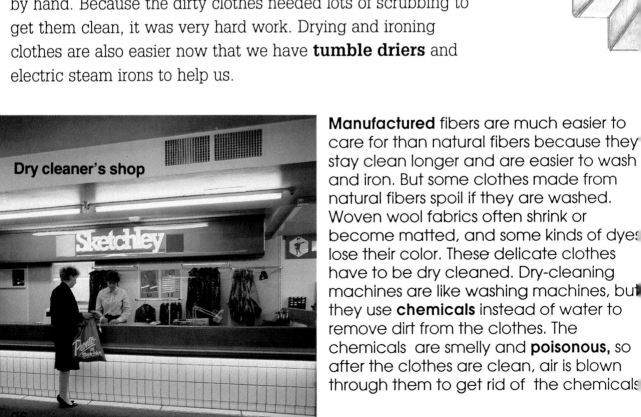

Dry cleaner's shop

Manufactured fibers are much easier to care for than natural fibers because they stay clean longer and are easier to wash and iron. But some clothes made from natural fibers spoil if they are washed. Woven wool fabrics often shrink or become matted, and some kinds of dyes lose their color. These delicate clothes have to be dry cleaned. Dry-cleaning machines are like washing machines, but they use **chemicals** instead of water to remove dirt from the clothes. The chemicals are smelly and **poisonous,** so after the clothes are clean, air is blown through them to get rid of the chemicals

TECHNOLOGY TO TRY

Keeping clothes clean

BEGIN HERE
Keeping our clothes clean is so much easier today than it was in the past. We have modern devices such as washing machines, tumble dryers and steam irons, and new types of soap powders and liquids. This project is about keeping clothes clean.

First of all, look at these two pages. Read **Look around** and the script for the television advertisement. On page 30 there are some investigations for you to try.

Finding out about something is called research. You are doing research into keeping clothes clean. Try to make notes about your research while you are doing it. They help you to remember later on.

Brand New Dribble

Scene: The inside of Mrs. Dirtyclothes' kitchen. There are piles of muddy clothes in a basket.

TV presenter:
Today I am visiting Mrs. Dirtyclothes, and I'm going to find out how she does her washing.

Mrs. Dirtyclothes:
Oh dear! Look at all these dirty clothes. My children have been playing in the mud again. I'll never get their clothes clean!

Look around
Next time you visit a supermarket have a look at the soap powders and liquids. How many different kinds can you find? If you read what it says on the outside, you will see that most of them seem to claim to be the best at keeping your clothes clean. Can this be right?

When you are watching television, listen carefully to any soap powder advertisements. How much do you believe? What do you think of the story, "Brand New Dribble"? It is an invented script for a television advertisement. You could make up you own advertisement.

Places to visit
Supermarkets
Launderettes
Your kitchen
Dry cleaners
Clothing stores

Words to know
You will come across some unusual words in your investigations. If you do not know the meaning of a word, you could try the Word List on page 31 of this book, or better still, use a dictionary. Here are some of the words you may need to look up:

detergent
soap

biological
washing powder

TV presenter:
Don't worry, Mrs. Dirtyclothes. Have you ever tried brand new, liquid, low-temperature Dribble?

Mrs. Dirtyclothes:
Well ... I ... er ... No. Does it work?

Sometime later, when Mrs. Dirtyclothes has finished her washing.

TV presenter:
Well, Mrs. Dirtyclothes, how did you do? Let's have a look at your washing.

Mrs. Dirtyclothes:
Oh, this is wonderful. Just look at these lovely white shirts. Oh, it's marvelous. From now on I'll only use low-temperature Dribble. It's much better than anything else I have ever used!

Detergent or plain water?

Find out whether a detergent is better for washing things than just plain water.

Take a clean rag and cut it in half. Wipe both halves along the bottom of your shoes to make them dirty. Try to make sure that they both get just as dirty as each other.

Wash one of the pieces in clean water and the other piece in water with a detergent in it. Use water that is just as hot as your hands can stand and wear rubber gloves.

Wash each piece for the same length of time. Rinse and dry the rags. Look carefully at them.

Which one is cleaner?

Brand X or Brand Y

Now try comparing two different detergents. Do the same experiment again but use two different soap powders or liquids.

Which one was the best?

Plain or biological?

Some detergents are specially made to wash out food stains. Do the same experiment again but use different kinds of dirt. Try tomato sauce, coffee, cooking oil, jam, or any other kind of food. Use an ordinary powder and a biological powder.

Do biological powders wash cleaner?

Cold or hot water?

So far you have used water in all your experiments. Some detergents work in cold water, too. Make up your own experiment to find out which detergents wash cleanest in cold water.

TAKE IT FROM HERE

If you look at the labels inside your clothes, you will find that many of them have washing instructions written on them. Sometimes these are made of little pictures. Can you find out what these pictures mean?

When a new kind of cloth is made, people in factories test it to find out the best way of keeping the clothes clean. They do experiments that are much the same as the ones you have done. When they have done the tests, they know what to write on the labels. These people are some of the technologists in the clothing industry. Can you find out more about this kind of work? would you like to be a technologist when you grow up?

Things you need

Various soap
   powders
Rubber gloves
Some rags
Your shoes
Cooking oil
Coffee
Tomato sauce
Jam

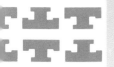

WORD LIST

automatic (page 26) Working by itself. An automatic washing machine washes, rinses and spin dries all by itself.
bar code (pages 22 and 24) A pattern of black lines that can be read by a computer. The bar code on a garment tells the computer what the garment is.
bobbin (page 12) A cylinder or cone made of wood, metal or plastic. Thread or yarn is wrapped around the bobbin so the thread or yarn won't get all tangled up.
chemicals (pages 10, 14 and 26) Substances used to change or make things.
crochet (page 16) A kind of knitting done with one needle with a hook on the end.
dyes (page 14) Liquids or powders used to color things.
fibers (pages 8, 10, 12 and 22) Short threads that can be spun together to make yarn. The threads may come from natural things, such as wool or cotton, or they may be made from oil or coal.
garment (page 22) A piece of clothing — a shirt is a garment, so is a sock.
Industrial Revolution (page 17) A time, about 150 years ago, when many people stopped working on farms and moved to towns or cities to work in factories.
lace-making (page 16) A way of making pretty net fabric patterned with holes. It is done by hand using many bobbins hanging from a frame.
light pen (page 24) An object shaped like a pen that can read patterns of light and dark, such as **bar codes**. It is used to send messages to a computer.
manufactured (pages 10 and 26) Man-made — made in a factory.
mule (page 13) A machine for spinning fibers into thread. It was invented by Samuel Compton in 1779 and is still used today.
poisonous (page 26) Something that can make you very ill or even kill you.
receipt (page 24) A piece of paper which shows that you have bought something.
safety helmet (page 7) A hard type of hat worn to protect your head at a building site.
shuttle (page 21) A type of bobbin in a sewing machine or weaving loom. It goes backwards and forwards, carrying the thread.
Stone Age (pages 8 and 20) A time, thousands of years ago, when people lived in caves and made simple tools out of stone.
tatting (page 16) A way of making a kind of knotted lace. It is done by hand using a shuttle.
tumble drier (page 26) A machine that dries clothes by tumbling them around in hot air.
warehouse (page 25) A place where things can be stored before they go to stores to be sold.
yarns (page 8) Long strands of thread that can be woven or knitted into fabrics.